Due to the some graphic events,
Reader discretion is advised.

ISBN 978-1-105-65299-8

Note from the author:

I began writing this series to share the stories and challenges faced by Emergency Medical Services workers. EMS is a stressful and rewarding career that is not for everyone. Many go to school, get their certifications, and start working on an ambulance for a fulltime municipal service, volunteer service, or private ambulance company. Many say they do it because they want to help people, they want to save lives, and they want to make a difference. That's why I started in this field, but that's not why I am here today. The truth is, you don't save that many lives, and the ones you do save will most likely forget about you in a month or maybe even a week. EMS professionals are the unsung heroes. We don't do it for the recognition, and we definitely don't do it for the money. The guy on the back of the garbage truck that picks up your trash every Thursday gets paid more than we do. We do it because no one else will, we do it because it is what we love to do. What you will read in this series are stories that are based on true events and people. Many details have been changed to conform to HIPPA laws and regulations. This series is dedicated to all the unsung heroes, the Artists in the Ambulance.

EMS Prayer

As I perform my duty oh lord,
Whatever be the call,
Help to guide and keep me safe,
From dangers big and small.

I want to serve and do my best,
No matter what the scene,
I pledge to keep my skills refined,
My judgment quick and keen.

This calling to give of myself,
Most do not understand,
But I stand ready all the time,
To help my fellow man.

To have the chance to help a child,
Restore his laugh with glee,
A word of thanks I might not hear,
But knowing is enough for me.

The praise of men is fine for some,
But I feel truly blessed,
That's you oh lord have chosen me,
To serve in EMS.

-Author Unknown

Chapter 1

Morning

It started out like any other day, but we knew something wasn't right. The sky was blue, a warm breeze was blowing, and birds were chirping, not a normal morning for the month of February in New England. As I entered the ambulance station to start my shift as an Emergency Medical Technician on Medic 7, I could already hear the tones going off and the fire dispatcher's voice coming from that big circle speaker in the ceiling.

"Medic 2, Rescue 2, and Engine 7, respond to 314 Montgomery for a two car motor vehicle collision with injury and entrapment". Then as I was punching in, another tone came over the intercom, "Medic 3 and Engine 6, on an echo response, respond to 100 Badger Road, a 75 year old male cardiac arrest". Already I knew it would be a busy day, but little did I know, it would be a day that I would never forget.

I found my partner already in the ambulance, doing the daily truck check. I picked up the check sheet and went around to the driver's side door. I opened the door and pulled the hood release handle to begin checking the fluids. After putting on a pair of gloves, I checked the oil, power steering fluid, and brake fluid, and found all were full, so I closed the hood.

In the front of the ambulance there were boxes of gloves, reflective vests for motor vehicle accidents, binoculars, and map books. I turned on the emergency lights, scene lights, headlights, and turn signals and walked around the truck to make sure they were all working and that there was no damage to the truck from the previous crew. The portable radios were in their chargers, and I took mine out and put it in my straps and handed my partner hers through the opening into the back.

My paramedic partner was already checking the medications, so I began to check the other equipment. In the outside compartments there were two back boards, and a scoop stretcher in one compartment. Extrication equipment, straps, and the trauma bag were all in another compartment. On the other side of the rig were the traction splints, and the infant car seat, still in its factory plastic wrapping since we had never used it yet. Back inside the truck I checked the jump kit. The jump kit was filled with an assortment of equipment found in the rest of the ambulance. There were scissors, pen lights, and bandages of all different sizes, band aids, burn sheets, and respiratory equipment. Next was the oxygen and medical air tanks. I found one of the portable oxygen tanks was low, so I took it to exchange it with a full one at the other end of the station.

As I passed by the kitchen on my way to where the other oxygen tanks were, I overheard some of the other Paramedics and EMTs on the shift talking.

"It sure doesn't feel right today" said Bill.

"I know, I was just thinking that too" Gerry said.

"What did the weather guy say on the news, 80 degrees today? This just isn't normal" Jerome chimed in from across the table.

"Something tells me it's going to be a busy day today" said Bob.

I put my empty tank down in the empty tank rack and picked up a full one and headed back to the truck. I passed the supervisor, Mack, in the hallway. As I got to the end of the hallway and was about to open the door back into the ambulance bay, I saw Mack passing the doorway to the kitchen. He paused and starred in at everyone. What came next was the normal Morning wakeup call for most of us.

"Why don't you all get off your lazy asses and do some chores around here. All this talk about nice spring days, what's next, you gonna start talking about picking flowers and petting cute fluffy bunnies? The floors need to be swept and mopped, the bathrooms need to be mopped and the toilets cleaned, and for the love of God somebody please put away the dishes that have been sitting in the sink since Monday!"

It was barely even 7:00am and Mack was already shouting.

Chapter 2

Mack

Mack was what you would want every shift supervisor to be. He always showed up with his black boots polished to well that you could probably shave using the reflection. His EMS pants were always clean, and starched, and pressed with military creases. Honestly! Who has their EMS pants pressed and starched? I mean, I keep my uniforms clean, now don't get me wrong, but I know that during the course of my day, I will be kneeling, bending, possibly even crawling in god knows what. This guy knows this too, but still every day he shows up the same way. Mack's white supervisor shirts were the same way too. Bleached white, starched, with military creases. His gold badge always glistened in the light. Serving Since 1989 read the plate above his name tag. Mack really has been doing this job a long time, and has seen it all.

Mack started about 23 years ago as an EMT. He took

the hazing and the jokes that were dished out, but it never got to him. Mack was always studying and reading everything he could, from cardiac dysrhythmias to digestion and even hair growth. Mack eventually went back to school and obtained his EMT- Intermediate license and continued his reading of medical journals, case studies, and even attended a few open autopsies. Mack always wanted to learn more about the human body, and one day, he was accepted into Paramedic School. Now that may not seem like much to a lot of people, but it is extremely hard work.

In Paramedic School, Mack had to memorize three different medication types every week for twenty four weeks leading up to a grand total of 72 medications. Now he had to not only memorize these medications, but he had to be able to differentiate between them and what they are used for and how they work. Dosages for adults and all the different dosages for kids, mechanism of action, contraindications, and routes for administration were just some of the things he needed to be ready to come up with quickly, and in an emergent situation. Mack also learned different techniques and skills like intubation, needle cricothyrotomy, and interousseous access were just a few of the new skills he learned. After almost three years of classroom lectures and internship time, Mack was a paramedic.

Mack was what many would call, a "trauma junkie". He liked trauma calls more than any difficulty breathing call or cardiac call that came in. Mack always said, "Trauma calls are easy, if its bleeding stop it, if its broken splint it, pop two IV's into them, then drive like hell to the trauma center. Simple!" Now don't get me wrong, Mack never liked to see anyone hurt, but this is what part of the job is. People get hurt, and they need a well-trained ambulance crew to pick them up, or what's left of them, and take them to the hospital where they can get patched up and hopefully make as close to a full recovery as possible. It wasn't that Mack disliked the regular old medical calls, he just preferred the trauma ones. Mack told me one time, "someone gets into an accident, bones are broken, organs are bruised and bleeding, they are having the worst day of their life and the worst day of their family's life. If I can be there to take care of them, keep them from getting more injured, treat them as best I can, and deliver them to the hospital doing a little better than when I found them, then I have done my job and that patient is better for it."

Mack wasn't what many refer to a lot of paramedics as a "paragod". What is a paragod? A paragod is a word used to describe that paramedic, either old or new, that thinks that every patient will die if they aren't the one caring for them, and that everyone who isn't a paramedic like them is just an ambulance driver. Mack wasn't that, by far. Mack always had time to answer your questions or sit and talk with you about a bad call, and he would never make you feel like he was better than you. Mack took the time to explain things so that you understand, and if you don't understand, he would explain it a different way. Mack never gave up and never got angry or frustrated. He was always there, and would put his supervisor work aside to help a fellow EMT or paramedic in need. Mack was my go-to-guy that I could go to whenever I had a problem or question.

Chapter 3

Medic 7

Now here on Medic 7, we don't usually use any of the emergency equipment on the truck. Sure we get dispatched to the local nursing homes from time to time for someone who may have fallen and broken their hip, but really Medic 7 is assigned as an interfacilty transfer truck. What's the meaning of that? We bring your grandma and grandpa home from the hospital. No, really we transport patients from one hospital to another, or to a nursing home or rehab facility. Here in Brownsville, Medic units 1-4 are the primary emergency response units dispatched by the county to respond to 911 calls in the area. Medic units 5-7 are used as interfaculty transport trucks, but do provide back up to the county should all the other ambulances be out on calls. With six other trucks ahead of us, we usually didn't get the opportunity to do those emergency calls. I can't remember the last emergency call we responded to other than the old man down the road who calls our ambulance dispatcher from time to time to have an ambulance come to help him find his dog. The saddest part is when you have to explain to him for the hundredth time that his poor old dog passed away many years ago. We sit with him and console him for a while, and then he sends us on our way. I haven't been back to his house ever since Engine 7's station dog had puppies and the guys brought one of the puppies to the old man to have. I think that was the best day of his life.

Here on Medic 7, we do a lot of transfers out of the local hospitals down to the big hospitals in Boston, Massachusetts, or up to Dartmouth Hitchcock Medical Center in Lebanon, New Hampshire. I've taken kids having allergic reactions to fava beans, teenagers who thought it would be cool to drink, drive, then park their car on the hood of a semi-truck on the highway, teen mothers in pre-term labor, and there's always "that guy" who thought it was a smart idea to turn the burners on his gas grill on with the top closed before lighting it. Yeah, we see it all. Medic 7 does them all. A lot of the calls we get are the psych transfers to the state hospital, the 70 year old lady who had a knee replacement and needs to go to the inpatient rehabilitation center, and the hospice patients who the hospital has made as comfortable as possible and now they are going home to die. Hospice calls, those are the ones that really get to me.

Medic 7 is a Ford E-350 box style ambulance. It doesn't look like any of the others in the station because it was purchased from a smaller company in the area that closed its doors due to funding. It's white with a wide blue stripe down the side and gold trim. It has a gold lettering decals on the sides and on the rear windows. It has strobe lights in the grill and on the light bar and LEDs along the sides and the back. All the other ambulances in the station are much newer, but our truck is still going strong. The inside is tight with new flooring recently put in the back during the last overhaul. As I completed my daily checklist, I tested the cardiac monitor and defibrillator and ensured it was in working order, as well as the IV pumps. Everything checked out ok and we were about to take our rig out to top off the fuel and wash the exterior when my partner and I noticed all the other trucks had left to go on calls. Just then, the portable radio I was carrying went off.

"Medic 7, Medic 7 go in service with county fire dispatch, call pending". My partner and I looked at each other.

"I guess we don't need fuel anyways" she laughed. I turned my radio channel over to county dispatch and told them we were in service. Just then the tones went off and we heard,

"Medic 7, Rescue 2, Truck 1, Engine 8, Car 1, respond for a fall with injury, 308 Haverhill Road".

We jumped in the ambulance, put on our seatbelts, turned on the lights, and away we went.

Chapter 4

The Pool Incident

While we were responding, we were updated by county fire dispatched over the radio.

"All companies responding to Haverhill Road, you have a 35 year old male, fell into an empty pool, not conscious, unknown if breathing."

I could tell by the dispatcher's voice, and the fact that they dispatched pretty much the whole world on this one call for a "fall", that this could be a bad call. Now, this could go one of two ways. We could get there and find the patient up walking around with some small bumps and bruises, or we would get there and find a patient suffering from a major trauma.

The last time I was dispatched for a cardiac arrest call, we got on the scene to find the patient stumbling around. After talking to him, we discovered he was drunk, and had called 911 because he was hungry. Knowing that this call could go either way, we always prepare for the worst.

When we arrived on the scene, my partner and I grabbed our stretcher, threw a backboard on it along with our cardiac monitor, jump bag, and oxygen equipment, and followed a young firefighter into the backyard where we found a swarm of firefighters in yellow coats standing around the edge of a pool. A woman was crying and hugging the fire chief on the back deck of the house. I walked up to the edge of the pool and noted a man who appeared to be in his mid-thirties lying on his back at the bottom of the deep end of the pool. Three firefighters were at the bottom of the pool as well in about seven inches of water that has a pink tinge to it. There was a small sump pump lying at the bottom near the patient. One of the firefighters was performing ventilations on the patient using a bag valve mask, while another was obtaining a blood pressure.

The captain from the engine company came over to us and stated, "His wife said he was trying to lower the sump pump down to remove the rest of the water to fix a hole in the liner when she saw him slip on the edge and fall in head first. She said he was face down in the water for about 7 minutes before we got here and got down to him. He wasn't breathing but does have a weak pulse".
My partner and I climbed down a ladder the firefighters had placed into the deep end, and began out assessment.

Everything they teach you and drill into your head in school helps in these situations. C-spine stabilized, check responsiveness, airway, breathing, circulation, look for obvious bleeding, then any obvious injuries, vital signs, then a thorough assessment, and package. I took in the scene for a few seconds. The man was lying in the deep end in the pinkish, green water. There were leaves and other debris strewn about in the water, and it had that nasty rotten vegetation smell. I noticed what appeared to be some teeth in the water as well. I picked up the pieces of white to get a better look at them. Yeah, they were teeth. I crouched down and began looking the patient over more closely. His nose was obviously broken, and blood was coming from it. There was blood and teeth in his mouth that the firefighters were trying to suction out every once in a while as they were still trying to ventilate him with the BVM. His trachea was midline and there was no jugular vein distention. My partner went back up the ladder. "I'm going to get set up and alert the trauma team, package him up and get him up here" she said.

The back of his neck felt funny, like something wasn't right.

"I think we have a step off" I shouted up to my partner. I assessed his shoulders and found deformity on the left side. A firefighter helped me apply a cervical collar to the patient's neck. I continued feeling for abnormalities. His chest was stable and intact, so were his arms. His abdomen was hard, not like it should be, and his hips just didn't feel right either. I went down his legs. There was an open fracture to his left lower leg that was bleeding. The right leg felt normal with no deformities. The firefighters above lowered down the backboard in the stokes basket. The firefighters in the bottom of the pool rolled the man on his side, I assessed his back finding no other signs of injury, and rolled him back onto the board. We secured him with all the straps, and the guys up top pulled him up.

I climbed up the ladder and found the firefighters had placed the patient strapped to the backboard onto my stretcher, and were securing him with the safety belts. One firefighter was still ventilating him with the BVM as they hurriedly rolled out of the backyard and to the ambulance. I grabbed the rest of our equipment and quickly made my way to the ambulance.

In the back of the truck I found my partner setting up the intubation kit. She had a 7.0 ET tube with stylet out and the laryngoscope ready. I began reassessing the patient. His left pupil was eight millimeters, and sluggish, the right pupil was four millimeters and sluggish as well. Blood was still coming from his nose, and his mouth was bleeding as well with multiple teeth missing. He had a large hematoma to the left side of his face. He was unresponsive with a Glasgow Coma Scale rating of 5 out of a possible 15. As they teach you in school, "GCS less than 8, need to intubate". He already had an oral airway in his mouth so we knew there was no gag reflex. My partner instructed the firefighter who was still ventilating with the BVM to increase the rate. She got her laryngoscope ready in her hand as the firefighter removed the BVM and oral airway. I suctioned the patient's mouth and she inserted the laryngoscope. I handed my partner the tube, she inserted it down into the patient's throat, inflated the cuff, and removed the stylet. The firefighter attached the capnography sensor to the end of the tube and then began ventilating again. I watched the numbers on the monitor, we were in. The patient was oxygenating well and went from an ashen color to a more pink color. I attached the monitor and automated blood pressure cuff. Once we had the patient on all of our monitoring equipment and the firefighter was taking care of the ventilations, my partner and I began looking for site to place IV's.

I looked on the right arm and my partner looked on the left. After applying the blue rubber tourniquet to his upper arm, I watched the vein inside his elbow begin to pop up.

"I've got one! " I exclaimed.

I took out a 16 gauge IV catheter and start kit, and set up a bag of normal saline. After wiping the site down with an alcohol pad, I took hold of the IV catheter and began to advance it into the vein. I watched blood run up the catheter and into the collection area known as the "flash chamber". My IV was in. I retracted the needle and advanced the catheter in at the same time. After releasing the tourniquet, I used my right hand to hold the catheter and occlude the vein so that blood wouldn't run out of the catheter. After attaching the IV tubing and making sure that the fluid was running in and there was no sign of infiltration, I secured the IV in place with a square tegaderm and some pieces of tape.

My partner then turned to me, "drive fast, just don't kill us".

I jumped up to the driver's seat and called county dispatch on the radio and told them we were transporting with a firefighter on board and engine following. I switched the radio over to the local hospital for my partner to patch in.

"Methodist Memorial this is county Medic 7 with status 1 traffic, how do you copy?" I heard her say.

"Go ahead Medic 7, this is Methodist Memorial" a voice answered through the speaker.

My partner responded, "This is Medic 7 inbound with a thirty- five year old male who suffered a fall into an empty swimming pool from a standing position. The pool was approximately twelve feet deep from the edge to the bottom. We are requesting trauma team activation. He is currently unresponsive with a Glasgow Come Scale score of five. Left pupil is dilated, right is constricted. Patient is intubated with positive lung sounds, two large bore IV's. Vital signs are pulse one thirty- six, blood pressure is seventy- six over forty, ventilations are being assisted with a pulse ox of eighty- eight percent on high flow oxygen, ventilations are assisted, breath sounds are present bilaterally but diminished in the bases. We will be there in ten minutes."

As we were driving down the road, I heard on the county fire dispatch radio, "Medic 4 and Engine 3, Delta response, a motor vehicle versus pedestrian, 31 South Central Street, reported to be a twelve year old male struck by a pickup truck."
What is going on today? The first nice day we have since October, and people are just getting hurt everywhere.

As we arrived at the hospital I noticed that there were five other ambulances from our service parked in the ambulance parking area already. It occurred to me that I had barely heard anyone go out for any of the usual difficulty breathing, check the person after a fall, or transport the drunk guy from the jail calls that we normally have. Everything thus far had pretty much been a trauma call except for the normal 7:00am cardiac arrest call which is normally for some old guy who died peacefully in his sleep the night before.

I got out of the driver's seat and went around to the back of the truck and opened the doors. My partner grabbed the monitor and jumped out, and the firefighter we had taken with us continued ventilating the patient. I carefully pulled the stretcher out and let the wheels go down gently. This was one of those times when I wished we didn't have the battery operated stretcher since it takes forever for the wheels to reach the ground. Once the wheels were down, my partner unlatched the stretcher from the safety hook, and we rolled into the emergency department. We were directed by a nurse to go to trauma room 1.

Trauma room 1 always amazed me by the state of the art equipment they had. A portable ultrasound cart, the new color glide scope for those hard intubations, built in x-ray equipment, and even a pneumatic tube system to send blood and other substances directly to the lab. We wheeled into the room that was crowded by about twenty doctors, nurses, x-ray techs, and respiratory therapists all dressed in blue gowns with those blue little booties, masks with plastic face shields, and gloves. We made our way through the sea of blue and positioned the stretcher next to the trauma bed and adjusted the height with the buttons. The good thing about the battery operated stretchers was no more back breaking lifting, except getting in and out of the truck. All you do is press a button and it goes up and down. We transferred our patient to the trauma bed and I took the stretcher and our equipment out of the room. When my partner was done giving her report of what happened, the patient's injuries, and our life saving interventions, we headed back towards the ambulance bay to clean up, gather supplies, and type up the patient care report.

Chapter 5

Partners

Even though I've only been in EMS for about six years now, I have had the joy, and the horror of working with many different partners. Your partner is the key to how your day will go. They are the person who you will work with anywhere from eight to sometimes twenty- four to forty- eight hours. Some people in this line of work see many partners during the course of their career. I have had partners who are new and green, old and experienced, and everything in between. I have seen partners make decisions that have ultimately led to the poor outcome of patients, and I have seen partners bring patients back from death's door step to the point where they are able to walk out of the hospital on their own, and return home to their families. Your partner can be your best friend, or your worst enemy.

My current partner Mary and I get along really well together. We can anticipate each other's moves to the point where we can just show up and do almost an entire call without having to even say a word. We just jump in and do what we have to do. Mary has been working in EMS for close to thirteen years now. She started out as an EMT- Basic, then EMT- Intermediate, and now Paramedic. Mary started working in the EMS field in Alabama on a county based municipal system. Mary is a state licensed and nationally certified IFT Medic. To be an IFT Paramedic, you have to be certified to a higher level than a normal paramedic. IFT Medics are certified to be able to care for pediatric patients in critical situations, able to maintain and regulate the infusion of blood products, and operate ventilators to assist patients that are intubated or have a tracheostomy and cannot breathe on their own. She always tells me in that thick southern drawl,

"I've done my share of emergency calls down south, I'm just fine doing transfers all day".

Most people who get into this field do it for the emergency calls, they do it go the lights and sirens, and the excitement, but not Mary.

I don't quite know how to explain my last partner. She got her EMT-Basic license from her Health Occupations class in high school, and started working for our company at the age of 20. She never stepped foot on an ambulance until her first day of orientation. I was her field training officer. We were working on a basic life support ambulance and our first call was a routine transfer from the local hospital to a nursing home. I figured that since she had passed her EMT- Basic course at the top of her class, and passed through the company's week long orientation, she was an ok EMT.

I drove over to the hospital and she backed me into the parking spot. We went upstairs to the floor, I got report from the nurse, transferred to patient to our stretcher, and then proceeded down to the ambulance. It was a two mile drive to the nursing home we were transporting this patient to, so I told he she could drive and I would take the report for this patient. I got in the back and she got into the driver's seat. Now it was near the end of winter in New England and after the winter we had, the snow banks were about three feet tall, and hard to see around the corners where it was piled up to about seven feet tall.

As we were driving down the road, I heard a loud thud. There was a loud scream and the truck lurched to a stop. The dementia patient we were transported was unaware that anything was going on and she just kept patting her stuffed cat she was holding as if it were real. I poked my head up through the "dog house" hole to the front of the truck and asked what happened. "I hit a kid!" my partner shouted and started to cry. "How the hell did you hit a kid?" I screamed. While she turned into a blubbering pile of mush in the front seat, I looked out the side window and saw a twelve year old girl standing there holding the side of her head. I reached up front, and grabbed the radio and call the dispatcher for two additional ambulances, and the police department, and told them we had been in an accident. It turned out that the girl had been walking on the top of the snow bank on her way home from school when she slipped and fell towards the roadway, right as our ambulance was coming by. She was hit in the side of the head by the right side view mirror, and all she had was a small bump and a story to tell about how she got hit by an ambulance. Two additional ambulances arrived as well as the shift supervisor. One truck took over our patient while the other transported the injured girl to the hospital. That was the last time I had that girl for a partner. She quit right after that call and now works as a cashier at the local supermarket.

Chapter 6

Decon

After scrubbing the stretcher for about thirty minutes, I gathered all the supplies that we used and headed out to decon the back of the truck while Mary finished her report. That was one of the reasons I was glad I was an Intermediate and not a Paramedic. I did not envy the report that she had to write. Everything we did from dispatch to drop off and every single intervention in between. When I got to the truck and opened the back doors, there was what seemed like an ocean of blood on the floor. There were empty packages from the IV supplies and intubation equipment on the bench seat, and the blood filled portable suction tubing as well as the inboard suction tubing was intertwined on the floor in front of the airway seat. I don't know many people who can deal with a scene like this. If the original scene of bloody teeth and clots strewing from the patient's mouth wasn't enough to make one vomit, the aftermath of the truck was. This was one of the reasons I wished I had gone back to school to become a Paramedic instead of stopping at Intermediate.

I started got in the truck and pulled it to an end of the ambulance bay where there was a slight incline and parked the truck facing upwards. The hospital always has a garden hose with about one hundred feet of hose located at the side of the emergency department for situations like this one. I picked up as much of the used, blood saturated equipment and packaging I could reach and stuffed it into a biohazard bag. I sprayed all the surfaces with a bleach and water mixture, and began spraying the floor with the hose. I let the red water run out the back and into the sewer drain. When the water was done running out the back, I grabbed the mop in the dirty utility room and began mopping the floor with the bleach and water solution. There was blood on the ceiling. How the hell did blood get on the ceiling? That's the question we always ask. Anytime a patient is bleeding, the blood almost always gets on the ceiling somehow. After grabbing a Clorox Wipe, I wiped the spot off the ceiling, and then using more wipes, scrubbed the bench seat, airway seat, action area, light switches, cabinet doors, stretcher locking handle, and door handles.

The IV kit was put back together, as well as the intubation kit, trauma bag, cardiac monitor, and backboard bag. When I was sure everything was clean and sanitized as much as it possibly can be in an ambulance, I went to see if Mary needed anything else. Mary was still finishing up her report, so I went to get a demographic sheet from the emergency department secretary. The face sheet has the patient's legal name, address, and insurance information. When I got back to the EMS room where Mary was still typing, I gave her the demographics sheet and called the county fire dispatcher on the phone to get the call times for the report. The dispatcher told me that Medic 7 was still needed for county 911 coverage as the other trucks were still out on calls. Medics 5 and 6 had gone out of service with the county because there were transfer calls to do while we were on our trauma call, and there were still a lot of 911 calls coming in. Today was not a normal day.

Chapter 7

11:00am Drunk

Mary and I made out way back out to the ambulance and called the dispatcher on the radio to say we were clear of the hospital and heading back to the station. "Medic 7 stand by, E.M.D. pending, Westchester Street" replied the county dispatched. E.M.D. pending means that the E911 center is taking an emergency medical dispatch and they are gathering information on the call. We started heading towards Westchester Street with traffic when the tone came over the radio, "Medic 7 respond to 185 Westchester Street on the first floor an intoxicated person with abdominal pain, alpha level." Alpha level is the response code for a response that doesn't require the use of lights or sirens. We made our way across the city with the normal flow of traffic and arrived at the location about twelve minutes after we were dispatched. We got out of the truck, grabbed the jump bag, the monitor, and the stretcher, and headed in the building.

Now 185 Westchester Street is what you would call a homeless shelter. The less fortunate citizens can go in for a place to stay, eat, and receive minor medical attention. Needless to say, the 911 trucks do a lot of calls out of here. We were met by a facility staff member who approached us with an anxious demeanor about her.

"Oh my, what took you so long, come quick, he's down the hallway here" the staff member said.

Before I could say anything back, the staff member hurried down the hallway and around the corner, and so Mary and I followed. We entered the infirmary area of the building where we found a man appearing to be around seventy years old sitting in a chair holding his stomach. The facility nurse turned to us and said, "He's drunk, and he says his belly hurts". This was a normal call for us. Many times a day the transfer trucks get called to 185 Westchester Street because they don't like it when the fire engine and police come as it makes a big scene and the staff feels that it gives a bad impression of their facility. The staff usually calls the ambulance dispatch center directly.

As I approached the man, I could already smell that very recognizable odor. It was the a mixture of booze, body odor, urine, and feces. From the first time you smell that odor, you always remember to stay upwind as often as possible. I asked him, "How can we help you today sir?" "My stomach is killin' me man", the man replied while tripping over his words. When I went to start my assessment, the man suddenly slumped over and stopped talking. Mary and I picked the man up out of the chair and laid him on the stretcher. Mary cut off the man's shirt as I prepared the monitor leads. Mary took the monitor leads and the twelve lead ECG harness and attached it to the man's chest as I got a non rebreather out and applied fifteen liter of oxygen to the man. Mary set up the monitor while I obtained a blood pressure. "One eighty over ninety" I said out loud.

Mary pressed the 12 Lead button and the monitor began spitting out paper. Mary tore off the strip of paper and began staring at it intently. "He's having the big one", Mary said excitedly and she handed me the printout. As I looked at the pink and white checkered paper with the black lines on it that the monitor had printed out, I noticed the tell-tale signs that they teach you in class. There was ST segment elevation and depression all over the place. I hooked up all the seatbelts on the stretcher to make sure our patient was secured, and we quickly made out way to the ambulance.

After loading the patient in the ambulance, he started coming to again. Mary and I each grabbed an arm and looked for IV sites. We both found good veins and inserted the largest appropriately sized IV catheters we could and hooked them up to IV bags. The man was still complaining that his stomach hurt, but was now stating that his chest and lower jaw hurt as well. I jumped in the driver's seat as Mary took out her drug box. She took out four baby aspirin and had the patient chew on them while she took out a bottle of nitroglycerin. As we were driving to the hospital, Mary gave the patient a nitroglycerin tablet under his tongue. After a few minutes the patient told Mary that his pain was getting a little better and then asked, "Do you have a sandwich?" The man seemed to be speaking better during the transport and wasn't acting like an intoxicated person would.

The only reason for this change that we could think of is that he was hypoxic from decreased oxygen levels and decreased perfusion. Mary called the hospital on the radio and told them what we had, and the patient's current vital signs. Then she transmitted the ECG that we obtained as well as other that she obtained during the transport.

When we pulled into Methodist Memorial Medical Center, the emergency department doctor met us in the entrance. He asked to look at the ECG strips we had and after a few seconds of examining them he said, "Cath lab two, now, don't stop". We proceeded with our patient down the hallway past the trauma rooms and through the double doors to the cardiac catheterization lab where there was another team of doctors and nurses waiting for us. This time instead of blue, it was a sea of green surgical scrubs and gowns and lead vests. We transferred the patient to the cath lab treatment table and helped the hospital staff hook him up to their monitoring equipment. I then took the stretcher out of the room and waited outside for Mary. "Seriously?" Mary said to me as she walked out of the room and closed the door.

"This is insane, what happened to the 'routine transfers' that we normally do?" Mary said as she walked down the hallway pulling the stretcher.

"Today apparently isn't a normal day" I replied pushing the stretcher behind her.

In the EMS room I called the county dispatcher again for our call times and was advised that our services were no longer needed and we were clear of services with the county for the time being. After completing the patient care report and restocking the ambulance with the supplies we used, we headed back out to our truck. Mary and I decided it was probably a good time to find something to eat, seeing as how it was already thirty minutes after noon.

Chapter 8

Lunch and Other Issues

The field of Emergency Medical Services has a lot of myths and superstitions surrounding it. For instance, if you say "Q" word (quiet), it is said that your day will be the most busiest day from hell. It is also a well-known fact that if you do not bring your own lunch with you from home, you probably won't get a chance to eat unless you take yourself out of service, which is highly frowned upon. Even uttering the word "lunch" can cause the tones and voices in the speakers to come on, and send you out to the ends of the earth. EMTs and paramedics often times will go the entire shift without eating more than a few crackers and maybe a small packet of peanut butter from the EMS room at the local hospital, or even maybe a sandwich from Dunkin Donuts. Speaking of Dunkin Donuts, coffee is the blood that runs through your veins in EMS. New people to the profession find that out the hard way very quickly. Early mornings that turn into late nights and back into early morning, sometimes without any sleep at all force you to change your liquid intake habits. Now I haven't heard of anyone starting an IV on themselves and trying to infuse the large coffee with regular cream and sugar, but I would not be surprised if someone did. Again, this job, this calling is not for the weak of soul. It takes a strong person to go from being fast asleep (if they even get the chance), to being wide awake and responding to a call where someone's life hangs in the balance between living and death.

As Mary and I drove down the road, we began the great debate that most crews that get a chance to eat have, where to go. There are so many choices to choose from. There are fast food places like McDonalds, Burger King, Subway, and Wendy's, or you can do Chinese, or pizza. As we were sitting there at the stop light going back and forth trying to decide where to go to get a bite to eat, I realized something. I hadn't gone to the bathroom since I woke up this morning seven hours earlier, and I needed to go. That's another issue with folks in EMS. Always use the bathroom when you first notice you need to. If not, you may not have the chance for a while especially if you get stuck on a long transfer. Another thing that you should never do is use the bathroom when you are next up in the rotation. There's not much worse in the world than settling down on the john to take a shit when the tone goes off for a cardiac arrest, because it's going to happen. Use the bathroom when you are restocking at the hospital or when you are last up, never first up.

Mary and finally decided on pizza, and when we were starting the battle over what place to go to for pizza, we heard a tone go over the radio and we paused to listen.

"Medic 1, Medic 2, Rescue 1, Engine 10, Truck 2, Car 1, respond to Highway one- seventy- one, in the area of mile marker twelve point eight for multiple motorcycles involved in a crash".

Then another tone went over and we heard the dispatcher say, "Medic 3 and Engine 2 respond to thirteen Jonathan Boulevard for a motor vehicle versus pedestrian".

Mary and I looked at each other and just shook our heads. We decided to pull into the nearest pizza place we could find. As we were approaching the local house of pizza we heard Mack's voice on the radio. Mack was on Medic 1 today so we listened in to see what was going on.

"Medic 1 to dispatch, I need two additional ambulances to this scene right away".

I looked at Mary, "here we go again" I said. Just then the ambulance dispatcher called us on the radio and asked us to sign on with county fire dispatch. Mary turned on the county fire mobile data terminal and selected the "In Service" icon. Right after that the tones came over the radio again.

"Medic 4 and Medic 7 respond to the call, Highway one-seventy- one in the area of mile marker twelve point eight to assist the companies on scene with multiple motorcycles involved in a crash, operate on channel three".

"County dispatch, Medic 7 is responding code three from downtown" I said over the radio.

We were off again.

Chapter 9

The Motorcyclist

The response to the scene seemed to take forever. As we made our way through the city, lights flashing, siren screaming, horn blaring, I couldn't help but to let my mind race as to what we would find when we got on scene. Four ambulances, they dispatched four ambulances to an accident on the highway. An accident involving multiple motorcycles. Most of the motorcycle involved incidents I've responded to have been sign offs. We get dispatched with what seems every fire truck they have available, and get on the scene to find the patient up walking around yelling at the guy who cut him off or what not. Most people who have been riding motorcycles for a while know how to "dump" their bike when faced with a possible collision, but apparently this was not one of those calls.

Why don't people pull over to the right and stop like they are supposed to? That has to be my biggest pet peeve about driving lights and sirens to an emergency or when transporting a critical patient. The law states you are supposed to pull over to the right and stop. Most people just stop where they are, pull to the left, and some will even try to race or outrun you. Then there are those select assholes who will try to follow the ambulance as you weave in and out of stopped traffic during rush hour while transporting a critical patient to Boston. They get right up to your rear bumped and follow you. I've had partners that could look out and read the newspaper headlines of the newspaper the person was reading while following us down the highway, that's how close they get. I honestly can't stand it. Fortunately for them, anyone who drives an emergency vehicle knows they must drive with "due regard" at all times. That means stopping at all red lights and stop signs and obeying all traffic laws with the exception that you are allowed to exceed the posted speed limit by ten miles per hour except in school zones. Yes there are the select emergency vehicle operators that blow through intersections where they have the red light, and drive 90 miles an hours to the hospital with a patient with a stubbed toe, but for the most part, we are professionals at what we do.

As we made our way onto the highway on ramp, I could see the stopped traffic. I mean stopped! There were people out of their vehicles walking around trying to see what was going on. Luckily we had a highway patrol officer in front of us trying to clear a path for us with his cruiser, but even he was having a hard time getting through. I jokingly suggested to my partner "why don't we just pull the truck over here, grab the gear, and walk, we might get there faster". She laughed and kept hitting the air horn button. The stopped traffic seemed to go on forever both in front and behind us. Sure enough there was some asshole trying to tailgate us as we cut through the traffic, but oh well, I had more important things to focus on at the time. We were going between five to fifteen miles per hour at times, and then stopping behind the one or two people not paying attention while trying to put on their make up or pick something out of their teeth looking in their mirror. We rounded a bend and I could see about a quarter mile up ahead the mass of red and blue flashing lights. Police cars, ambulances, fire trucks of all typed littered the three lane highway.

As we approached the scene, I called on the radio, "Medic 7 to Highway one- seventy-one command, we are approaching the scene. Where would you like us?"
The deputy fire chief answered, "Command to Medic 7, come up in front of Engine 10, you have a status 1 patient".
A status one patient, of course. Why couldn't we have a sign off or something silly like a simple laceration or something of that sort? A status one patient means that the patient has injuries that are extremely life threatening. As we pulled up in front of Engine ten, I looked out the window and saw the scene.

There were five motorcycles that had driven off the highway. One person was already covered with a sheet off to the right, there were about thirty motorcyclists standing around their motorcycles just beyond the scene. Ambulance and fire crews were working on four other patients that were lying on the ground. We Engine ten's crew just down an embankment performing CPR on one person while trying to roll them onto a backboard. Mary and I got out of our ambulance and put our reflective vests on. Just then Mack ran up to us. Mack had a look on his face I had never seen before. He had blood streaking his white button down shirt and some other greenish color substance that I don't even want to think about what it could have been.

"I need you two to grab your gear and help Engine ten down there, they have a traumatic arrest. The patient was conscious when they got to him and just before you guys showed up he went into full arrest" Mack told us as he hurriedly made his way to his supervisor truck. Mary and I grabbed our cardiac monitor/defibrillator and drug box and made our way down the steep grade to the patient below.

The patient was laying on a backboard with a cervical collar already on with one firefighter doing chest compressions and another doing ventilations with a BVM. Mary quickly set up the defibrillator pads and looked at the monitor.

"V fib! Everybody clear!" Mary exclaimed.

She pressed the charge button on the defibrillator and it began to make a tone that steadily increased in pitch until it began beeping and a little red light started flashing. Zap! The firefighter quickly resumed chest compressions as I began looking for an IV site. The patient had multiple obvious broken bones and deep lacerations that blood was oozing from.

"I'm not finding anything here" I said to Mary.

"IO him then if you can't get a vein", Mary replied.

The IO, or interousseous access drill, is pretty much a specialized medical version of a battery operated drill with a hollow drill bit that you leave in a large bone so that you can infuse IV fluids or medications when you can't establish an IV. I grabbed the Easy IO out of the drug box and had another firefighter set up the IV fluid. The firefighters had already cut away the patient's clothes so that we could expose his body to find injuries. I measured two fingers down, and two fingers in from the patient's left knee. I prepped the area with alcohol swabs and pulled out the interousseous needle and attached it to the tip of the Easy IO drill. The needle went in smooth. After putting throwing the drill portion back in the drug box, I unscrewed the needle portion of the interousseous access device, threw if in the sharps container, and attached a syringe. I drew back on the syringe to make sure I was in the marrow, and then flushed saline through to make sure it was clear.

"We have a line" I exclaimed to Mary.
It was my first time doing an IO.

We began infusing saline solution using a pressure bag and administered a dose of epinephrine to help get the patient's heart going again. Mary pulled out the intubation kit, the second time today, and prepared to intubate the patient. Mary set up the endotracheal tube and laryngoscope and inserted the tube into the patient's mouth and down their throat into the trachea.

"I think we're in"Mary said as a firefighter hooked the BVM up to the capnography sensor and the tube. I took out my stethoscope and listened for lung sounds.

"We have lung sounds in all fields" I said.

"There's a good waveform on the capnography monitor" said the firefighter doing ventilations.

"Hold up a minute" Mary said to the firefighter doing compressions. She glanced at the monitor which showed what we would refer to as a bradycardic, or slow heart rhythm.

"Check for a pulse" Mary instructed.

"I've got a weak femoral and carotid pulse" a firefighter said.

"Let's go!" Mary yelled.

I grabbed our drug box as the firefighters loaded the patient into a stokes basket with the monitor and began the trek up the embankment.

We secured the patient on the backboard on the stretcher and loaded him into the ambulance. Mary and I jumped in along with two firefighters. We had another firefighter jump in the front to drive us to the hospital. As we started rolling down the road, the monitor began making an alarming noise.

"V fib again! Everybody clear" exclaimed Mary. She pressed the charge button and the monitor built up the energy like before.

"Everybody make sure you are clear!" she yelled again.

"We are" said one of the firefighters.

Zap!

"Start compressions again" Mary said.

One firefighter began compressing the man's chest again. I took out another dose of epinephrine from the drug box and administered it through one of the ports on the IO line we had.

"Epi's in" I said to Mary.

We continued chest compressions, defibrillations, and doses of epinephrine all the way to the hospital. Mary called into the hospital on the radio, "Methodist Memorial, Methodist Memorial, this is Medic 7 with status 1 traffic", she said.

"This is Methodist, go ahead", a voice replied back.

"We have an approximately fifty year old male involved in a motorcycle crash versus tree. Patient was responsive on the scene when fire got to him, but then went into cardiac arrest. We were able to get a pulse and blood pressure back after five minutes of CPR and one dose of epi, but he coded again about eight minutes ago. Delivered a total of three shocks since then with no conversion. We have an IO line established and patent, currently asystole on the monitor. We are four minutes out, any questions?"

"No questions, see you when you get here, trauma one on arrival, Methodist Memorial clear" a voice answered back again.

We pulled into the hospital and rolled into trauma 1 again to the sea of doctors and nurses covered in blue gowns, masks, and gloves. After the patient was transferred to the trauma bed, I rolled the stretcher out of the room and thanked the firefighters for helping, then went back into the room. They worked on the man for what seemed like hours. The trauma surgeon began performing procedures on the patient hat I had only seen on TV shows, and after what seemed like hours, but was actually twenty- five minutes, the trauma surgeon paused.

"Stop CPR" the trauma surgeon said as he looked at the cardiac monitor. The cardiac monitor just kept displaying three straight lines across the screen.

"I still see asystole on the monitor, can anyone find a pulse?" the trauma surgeon asked.
Two nurses began feeling for pulses at different places on the man's body where you would normally find a pulse, and another doctor used his stethoscope to listen to the man's chest for a heartbeat.

"Nothing" said a nurse.

"I don't hear anything" the doctor with the stethoscope said. The trauma surgeon removed his mask and cap.

"Time of death, 13:47, thank you everyone" he said.
I went to go decon the truck and replace the equipment we used while Mary went to write yet another report.

Mack had arrived at the hospital to see how everyone was doing. Mack met me at the ER doors. I asked him,

"What happened Mack? What caused it?"
Mack said that the group of motorcycles was riding to raise money for the American Cancer Association. Apparently a driver in a car tried to cut through the group of riders to try and exit the highway, when a couple riders lost control, and caused a chain reaction crash. In total, two riders had died, two others were being flown by medical helicopters to Boston hospitals with critical, life threatening injuries, and four others were being treated at local hospitals for bumps and road rash. The driver of the car was caught and arrested. He was charged with aggravated DUI, with more charges pending.

Chapter 10

Dispatchers, friends or foes?

When I went out to put the equipment away and clean up the mess we made, I called the county dispatcher on the phone again for the call times. The county dispatcher said they had enough ambulances in service and that our services were no longer needed. After I hung up the phone, I called our ambulance dispatcher over the radio and told them we were clear of service with the county.

"Received Medic 7, be advised you have a discharge coming out of Methodist Memorial Medical Center traveling to the Blue Haven Nursing Home at 14:15, I'll show you on scene" the dispatcher replied back.

"Received" I said and then dropped the microphone on the seat. Looks like no Lunch for us today. I quickly cleaned up the truck, wiped down all the hard surfaces with disinfectant, and put away all the equipment before walking back into the hospital with the stretcher.

Now I know it's not the dispatcher's fault, but they are always the ones that get blamed. Why can't they give this call to another truck? Why can't they wait until one of the basic life support trucks is clear in the next few minutes and have them do this call? These are questions that always pass through your mind at a time like this. I have worked in our dispatch office before, and I know what it's like for them in there too. On the road you are just one truck worried about yourself and your partner. In the dispatch office, you are worried about up 20- 30 ambulances and wheelchair vans, and the seventy transfer calls that are already pre booked for the day. It's a stressful job that requires constant multi-tasking. Answering phones, booking calls, dealing with angry facilities and ambulance crews, emergencies, and truck breakdowns are just some of the things dispatchers have to deal with every day, if not every minute.

Rule number one of this business, don't piss off your dispatcher. Rule number two is, when in doubt, refer to rule number one. The worst thing an ambulance crew can do is to give an attitude to the dispatcher over the phone or radio, or try to dodge a call. Your dispatcher can be your best friend or your worst enemy. I've had days where I have barely done any calls all day because the crew on Medic 6 made the dispatcher angry so they kept getting call after call. The best thing you can do in this career field if you want to keep your job and your sanity for as long as possible is to come in, check your truck, do your job, keep your head low, smile, thank people, and don't piss of the dispatchers. I promise you your job will go so much better with your dispatcher on your side.

Chapter 11

Routine Transfers

Finally, a "routine transfer". This is what our truck is for. A normal BLS transfer is just what we need with a day like today. I uploaded the call information to our laptop computer and opened up the information. Methodist Memorial 5th floor, Mrs. Smith in room 594, bed number 2, being discharged after hip surgery. After remaking the sheets and putting more blankets on the stretcher, we made our way to the elevator and went up the 5th floor. When the elevator doors opened, nurses and nursing assistants were all scurrying around doing their daily duties and answering call bells. We walked over to the unit coordinator's desk at the nurse's station, which was empty, and waited to be helped.

After trying to get someone's attention for about ten minutes, the unit secretary finally came back and asked if she could help us. I told the secretary,

"Yes, we are here for Mrs. Smith".

"Oh yes, I'm just getting her paperwork together, her nurse is Jessica" replied the secretary.

I took the packet of paperwork and started my report. I began entering all of the patient's information consisting of mailing address, home address, and health insurance, into our laptop computer. The most important part of the patient's demographic information seems to be the insurance information. Without it, we don't get paid. Pretty much everything in the medical field is paid for in one way or another by insurance or Medicare, but that's a story for another day. An ambulance ride can cost anywhere from $500 up to over $3500.

Some patients unfortunately get stuck with a bill sometimes. Why do they get stuck with a bill? Insurance companies require the transport to be medically necessary. Most emergency 911 calls are automatically covered, but non-emergency transfers are not unless medical necessity for ambulance transportation is proven. In private EMS, medical necessity is drilled into your head over and over again. The most important piece of paper we are required to obtain on transfers is the physician's certification for ambulance transport form. This is a legal document that is signed by a medical professional who states that an ambulance is required for transport. The insurance companies check the physician's certification form against the ambulance report that the attendant writes and if they find inconsistencies, they deny the claim and the patient gets the bill.

The nurse, Jessica, finally came after fifteen more minutes of waiting and gave us report on the patient. The patient, Mrs. Smith, had come in to have her left hip replaced due to a long history of osteoarthritis, and is going to the local rehabilitation center to recover more before going home. We found Mrs. Smith in her bed and moved her to the stretcher smoothly by sliding her on the sheet she was laying on, and secured her with the seatbelts. The transport went without incident and we delivered Mrs. Smith to the rehabilitation facility. Maybe now we would be able to get lunch, or at this point, dinner.

Chapter 12

Calm Before the Storm

We finally settled on the first place we came across to get food. A couple slices of pizza from the local house of pizza was on the menu for today, and with only about four more hours left in our twelve hour shift, we headed back to the station to eat what we could before the next call. The county fire radio was still chirping away.

"Medic 2, assist the police with an intoxicated person, the corner of Main and Summer streets, police are on scene".

"Medic 3 and Engine 7, respond to 47 Kennedy Way for a pedestrian whose foot was run over in the parking lot".
We finally made it back to the station.

At the station, Medic 4 was still in the bay and the crew was sitting in the kitchen area at the table trying to type their reports. Writing reports on laptop computers has its pros and cons. It sure beats the old fashioned hand written reports and is usually easier and quicker to get reports written, but with a recent software update, our report writing program seems to be causing us to have to stay extra late to complete reports. It usually takes anywhere from twenty to forty minutes per report on the laptop, and when you have ten calls in a twelve hour shift, you quickly get behind in your paperwork.

Transfer trucks usually have an easier time with reports as you usually have more time in between calls as most are pre scheduled ahead of time. Today has been an extremely busy day, and combined with this new software update, according to the Medic 4 crew, all the crews are behind in their reports. Looks like some people could be stuck here until close to midnight trying to finish their paperwork. Just then, the tones went off over the station speakers again.

"Medic 4, Truck 5, Boat 1, Car 1, respond to the area of the First Street bridge for a jumper. Police are also responding. Boat 1 deploy up river at the boat launch, time out 17:34". I watched the Medic 4 crew get up grumbling and walk out to the ambulance bay. I walked out behind them to get a breath of fresh air as I ate my slice of pizza, and noticed that it was just us and Medic 5 in the station.

I have never seen a day as crazy as today. It never fails though, the first warm day after winter brings everyone out of their homes and into the streets. During a normal twelve hour shift, we do about four or five non emergent transfers to and from nursing homes and rehabilitation centers, with an occasional trip to Boston. Today has been a very strange day full of innocent people getting hurt. I went back inside and sat down at the kitchen table with Mary and the Medic 5 crew. Don the paramedic on Medic 5 was talking about a cardiac patient that they had just taken down to Massachusetts General Hospital. As we were looking over the EKG strip that Don had printed out from the patient, the station phone rang. I picked up the phone and before I could say hello, the ambulance dispatcher said,

"I need Medic 5 and Medic 7 to sign on with county fire right away, they have multiple calls pending".

Chapter 13

Shots Fired

I keyed up my portable radio as I made my way to the ambulance.

"County dispatch, Medic 7 is in service".

The county fire dispatched answered me in a voice that made everyone realize that something bad was happening.

"Medic 7 and Engine 8, respond to 180 Oak Street for a shooting".

I began running to the ambulance. They always teach you in EMT school, "don't run", and "it's not your emergency", but I knew that something was different about this call. This was also my first call for a shooting, or any violent crime for that matter. I have done the falls, the car crashes, overdoses, and emergent transfers. I have even transported patients that have been shot and stabbed to other hospitals, but I have never been the primary truck responding to the scene of the incident. As I got in the ambulance and my partner opened the garage door, I heard Mack sign on the radio that he was also responding. When Mary got in the ambulance with the tear sheet from the dispatch printer, she was looking at it with a look of horror on her face. She turned to me and said, "It's a cop".

We began racing down the road to the other side of the city. The traffic was parting for me like the Red Sea for Moses for once. While driving down the road, thoughts were racing through my mind. We were responding to the scene when a police officer was shot. This wasn't a call that many of the 911 trucks go to for gang members who shoot each other, this was a cop. Everything that was buried in the back of my mind about ballistics trauma came flowing back to the front. ABC's, control the bleeding, assess for entry and exit wounds, spinal immobilization, and rapid transport. I pressed the gas pedal harder hoping we would get there in time.

We drove up hills, down alleys, under bridges, around corners, and through intersections, only slowing or stopping to make sure that it was safe to proceed. I knew we needed to get there fast, but I had to make sure we got there in one piece. I heard Engine 8 on the radio sign off on scene. We continued to race down the city streets. There were police cars coming off side streets and approaching behind us. I couldn't count how many there were. I turned onto Oak Street and saw the scene.

It was a dense residential neighborhood with multifamily houses on both sides of the street. A large crowd had already gathered on both sides of the street. I saw Engine 8 and Mack's supervisor truck. There were about twenty police cruisers, some in the street, on lawns, and the sidewalk. There were state troopers, highway patrolmen, detectives, and street officers everywhere. Some had their guns drawn in their hands; others had assault rifles and shot guns. Other police cars kept showing up. Some officers were trying to push the crowds or onlookers back as others were already streaming yellow crime scene tape around street signs and telephone poles. I pulled the ambulance up to where an officer was waving his hand frantically at us.

Chapter 14

Officer Down

Mary and I went to the back of the ambulance and grabbed the jump bag and trauma bag and followed an officer around the corner. Everything seemed to go in slow motion at this point. There was a sergeant shouting to some other officers to push the crowd back farther and detectives talking on their radios requesting the major crimes unit to the scene. As I made my way through the swarm of officers, I noticed shell casings on the sidewalk and in the street being guarded by other officers. Some officers were crying while others had looks of horror on their face. Their brother had been shot, and no one knew how bad it was yet. No one knew if he would live or die.

It seemed like hours had passed by even though it was only a few seconds until we came to him. There he was lying on his back, blood flowing from underneath him, down the sidewalk, and into the street. Bright red blood. The firefighters from

Engine 8 were already working on him. They had applied a non rebreather oxygen mask to his face. His face, I will never forget it. His face looked white, literally as white as a piece of paper. His eyes were staring straight up into the darkening sky as the sun began to disappear behind the clouds that began rolling in. The gold badge on his uniform shirt glistened in the remaining sunlight. Then he screamed out in pain.

The scream woke me from my trance and I crouched down and began my assessment. A firefighter was cutting away his clothing. There were found two entrance wounds to the right upper quadrant of his stomach as well as another entrance wound to his right thigh. There was blood squirting from his right thigh. Mack grabbed a tourniquet from the jump bag and wrapped it around the injured officer's right leg above the bullet wound. Mack then instructed a firefighter, "hold pressure on this, and don't let go". I looked at another firefighter and said, "Go back to the ambulance and grab a backboard and the stretcher." I found an exit wound on the back side of the right thigh as well as another entrance wound above his right knee. Mary was applying trauma dressings to his stomach already. I glanced up and noticed heavily armored officers carrying automatic rifles had arrived as well as more and more officers.

We rolled the injured officer onto his left side and found an exit wound to his back. We applied more trauma dressings to his back, and rolled him onto a backboard and strapped him down. He was losing a lot of blood. We picked the officer up on the backboard and placed him gently on the stretcher, and made our way to the ambulance. Once again things seemed to move in slow motion as we moved towards the ambulance at almost a run. The swarm of police officers around us stopped what they were doing and made a pathway for us leading up to the ambulance. From the crowd of people that was growing larger and larger we could hear people saying "oh my God", "I hope he makes it", "will he be ok?" As the firefighters were loading the injured officer into the ambulance, Mary and I jumped in and set up our equipment.

Mack went to the officer's right side and Mary sat on the bench seat and began looking for an IV site on his left. The swarm of officers had gathered at the back doors of the ambulance and I could feel that each and every one of their eyes was on us. Mary and Mack each found veins to put the largest IV's in that they could. I noticed that the blood had stopped coming from the officer's wounds. A firefighter and I hooked up the cardiac monitor and automatic blood pressure cuff and began reassessing the officer's vital signs. I noticed that the officer's blood pressure had dropped significantly from the vital signs that the firefighters obtained when they first got on scene. Mary and Mack opened their IV lines up and began running saline fluid into the officer to try and raise his blood pressure. Every once in a while the officer would cry out, a noise that sent chills through my body. Mack instructed me to get in the driver's seat and get ready to head to the hospital. I jumped out the side door and threw my bloody gloves on the floor of the ambulance before closing the door. Quickly, I ran around to the driver's seat and jumped in and buckled up.

Mack told me, "Let's go, now!" and away we went. There were two police cruisers in front of us and three behind us as we flew down the street. Other cruisers were blocking off large intersections as we passed through. We screamed through the city to Methodist Memorial Medical Center. I could hear Mary in the back talking to the injured officer.

"Squeeze my hands!" she was shouting. "Good!" she said, "can you wiggle your toes now? Excellent!"
Now I can tell you that the transport time was not long at all. Mack had already called into the hospital and the trauma surgeons were actually waiting for us outside when we pulled in.

Chapter 15

The Aftermath

We rolled the injured officer into trauma 1 and moved him onto the trauma bed. The sea of blue clad doctors and nurses were there as usual. One of the nurses turned to the group of officers that had followed us in and said, "I'm sorry, you can't be in here right now, we will update you in a few minutes". Mary and I removed our stretcher while Mack gave the report to the doctors and nurses. A portable x-ray machine was being rolled in as we exited the trauma room, and I could hear Mack talking.

"Twenty-eight year old male police officer shot with a small caliber hand gun from what we can count four times. He has two entrance wounds to the abdomen with one exit wound to his posterior, another entrance and exit to the right thigh, and another entrance wound above his right knee. There was major blood loss on the scene and a tourniquet was applied to the right leg due to an arterial bleed. He decompensated in the field and we were able to bolus him with about fifteen hundred of saline. Two sixteen gauge IV's in place with saline wide open. Last blood pressure was 90/50, heart rate was 136, and he is breathing on his own at a rate of 20 on high flow oxygen by non rebreather".

I walked out to the ambulance and just sat down on the back step. Why did I come to work today? I knew this morning when I woke up it would be a life changing day. We all knew this morning that it was going to be a strange day. We had done all that we could do and hopefully he will make it. There were about forty police officers both in and out of uniform standing outside and inside in the hallway. Some were on cell phones, some were crying, some were hugging each other. I just sat there reflecting on everything that had happened that day. Just then two police cruisers flew into the ambulance bay with their lights and sirens on. A woman jumped out of the passenger seat of one of them and was led into the ER. I assumed she was the officer's wife. After a few minutes, the trauma surgeon came outside to address the mass of officers.

"He's lost a lot of blood, but we are transfusing him right now and heading up to surgery. He's in good hands and I believe he will pull through" the trauma surgeon said.

I began cleaning up the truck with Mack and Mary. Some detectives came around and asked everyone who was in the ambulance questions about what happened and what they saw. After I had cleaned the blood and from the truck and the stretcher, I looked at my watch. It was 7:30pm, only a half hour left in the shift. Mack called our dispatcher and told them we were remaining out of service for the remainder of the shift in case the officer's injuries were too severe that Methodist Memorial had to ship him out to Boston. We remained at the hospital with the crown of officers that had now grown to a mixture of about 60 city police officers both on and off duty, state troopers, sheriffs, highway patrol officers, and numerous representatives from other agencies waiting to hear word on how their brother was doing.

We had received word that a neighboring town had caught the shooter after a high-speed chase. A crowd of people had also gathered on the opposite side of the ambulance bay after hearing word about the shooting of the police officer on the news. After a few hours, the same trauma surgeon returned to the ambulance entrance doorway.

"There were some complications during surgery, but we were able to control the bleeding and I believe he is going to make a full recovery" the surgeon stated.
The officers outside began clapping and the officer's wife went up and hugged the trauma surgeon. After the day we had today, it was worth hanging around to find out that the officer was going to be ok.

Chapter 16

Drawing Conclusions

I left work that night and went home and hugged my wife for a long time. It had been a day full of drama and trauma. There was a lot of good medicine being done that day and I was glad that I was able to be a part of it.

-The man who fell into the pool was monitored in the intensive care unit and intubated for two weeks before he eventually succumbed to his injuries. The doctors said he suffered a massive brain bleed. His family decided to donate his organs to people in need of organ transplants. Mary and I actually transported the patient that received the man's heart to the local rehabilitation center a few weeks later.

-The man from the homeless shelter underwent a cardiac catheterization that day. The last I heard, he had been reunited with his son afterwards and had moved in with him somewhere in the Midwest.

-The family of the motorcyclist that died never really was able to get any closure, and we began transporting his widowed wife to the psychiatric hospital on a regular basis for multiple suicide attempts until the day she unfortunately succeeded and was buried next to her husband. The driver that caused the incident that day was found guilty on multiple chargers including DWI and vehicular homicide and was sentenced to three life sentences.

-Mrs. Smith remembered her favorite ambulance crew and began bringing in fresh baked cookies to the ambulance station for everyone. She unfortunately suffered a fall one day at home and broke her pelvis. She underwent surgery to repair it, but was never able to walk again. She passed away in a nursing home from pneumonia and sepsis.

-The police officer that was shot was eventually discharged home from the hospital. He unfortunately suffers from post-traumatic stress disorder now, and has taken a desk job at the police department until he can be cleared to return to work on the streets again. I have spoken with him recently, and he says he is slowly but surely making progress and hopes to be back on patrol within a few months. He always says how grateful he is to the ones who responded that day and saved his life.

The life of an EMT is full of stress and drama. EMS can take a fun, down to earth person and turn them into a cynical prick. Those who continue to work in EMS can make a lasting career out of it, as long as they don't suffer a severe debilitating back injury as Mack did a few months ago while trying to carry a patient on a stair chair down from a four story walk up. I haven't seen Mack in a while, but the last time I talked to him, he was lying in a bed in the local rehabilitation center, trying not to think about the pain. Yes I still believe that EMS is a very rewarding career. You do get the chance to make a difference in people's lives, and on some occasions, maybe even save a life.

The life expectancy of an EMT or paramedic is reduced up to 10 to 20 years. The stress and fast food gets to a lot of people after a while and can cause them to have a heart attack or stroke at earlier ages. Some even lose their families sometimes. The long hours and stress that you try not to take home with you, can wreak havoc on a marriage or relationship. Luckily there is another family that will always be there for you. The ones who know what you are going through. The ones who walk with you through the mud and blood on a daily basis. Those whose soles of their boots will forever test positive for MRSA, VRE, and God knows what else. Those people who have been there, seen it, and done it, the artists in the ambulance.

So take my advice. Wear your seatbelt, don't drink and drive, and please for the love of Gods turn the handles to the pots on the stove inward so that children can't knock them over. Don't pat strange dogs that you don't know or don't know you. Watch your step, don't stick things in electrical outlets that aren't meant to go in them. If you are allergic to peanuts, don't eat a peanut butter and jelly sandwich. When someone cuts you off on the highway, don't chase after them. Instead take a deep breath, and think about your kids or your family. Look twice before changing lanes. Brake for moose, trust me on this one. Don't eat yellow snow, its either dog or human urine and there's really no definitive way to find out. My last piece of advice is really a request. Please, thank an EMT or Paramedic. We put up with a lot of bull shit and stress, and sometimes it's that "thank you for what you do" from a stranger than can change their whole outlook on the day.

www.ingramcontent.com/pod-product-compliance
Lightning Source LLC
Chambersburg PA
CBHW021912170526
45157CB00005B/2051